Access scaffolding

C. J. Wilshere

Thomas Telford Ltd

1981

CONTENTS

Other ICE Works Construction Guides available
Pile Driving, W. A. Dawson
Earthworks, P. C. Horner

ISBN 0 7277 0090 1

Published by Thomas Telford Ltd, Telford House, PO Box 101, 26–34 Old Street, London EC1P 1JH

Typeset, printed and bound by David Green (Printers) Ltd, Kettering, Northamptonshire

1. INTRODUCTION

Purpose

For both new construction and maintenance, it is necessary for the operative to have convenient access to the point where he has to carry out his work. There will be some form of platform for him to stand on, and also probably to place materials on, together with a means of access, either horizontally or vertically or both. The risk of injuring himself or those below must be reduced as far as possible, but in addition the quality of work carried out from a good access platform will be improved.

Temporary constructions fulfilling these requirements have been in use for a long time. At the turn of the century they were of timber, with vertical round poles connected together with ropes by scaffolders. These men had learnt their skill through working on the job, and such scaffolds were usually very well made but not often high. In the early part of the century the use of steel tubes and fittings commenced, and this has gone from strength to strength, now being in use all over the world, although the greatest usage is probably in the UK. For the last 50 years scaffolds involving frames of various sorts have been in use as well and today proprietary scaffolding forms the major part of straightforward access scaffolds. This guide is limited to the application of metal scaffolding although timber and particularly bamboo are very effectively used in some countries.

Types of scaffold

The main demand for scaffolding is for a working deck parallel to the face of a structure. This is usually provided by an independent scaffold, described on p.20. Platforms are provided at various levels supported by pairs of vertical tubes. When constructing brickwork a putlog scaffold (p. 23) may be used, with one line of standards, using the building itself to support the inside of the platform. Scaffolds which serve an area are called birdcage scaffolds (p. 25). A scaffold can be hung from outriggers high up the building, usually on the roof, and may be moved vertically on ropes to the desired level. This is known as a suspended scaffold (p. 25). Towers with castors to enable them to be moved are frequently used, and are dealt with on p. 25.

An important variation is the addition of sheeting to provide protection from the weather, or to retain the debris of construction within the scaffold itself. If nets are needed to catch falling debris — or operatives — reference should be made to BSCP 93.[1]

Organization of industry

Most scaffolding is provided by specialist scaffold contractors, who supply and erect equipment, usually on a hire basis. They may where necessary adapt and eventually dismantle it. Contractors may hire equipment but erect it themselves, but most own at least a small amount of scaffolding. Similarly, organizations with their own maintenance department will frequently own some equipment.

Where scaffold is to be obtained by means of a sub-contract, it is important that the user explains his requirements clearly to the potential contractor. The BSI Draft for Development for Performance Requirements[2] sets down many of the technical requirements for such a contract.

Table 1. Information on typical rule of thumb scaffolds

Distributed loads on platforms	75kg/m², inspection and very light duty	150kg/m², light duty	200kg/m², general purpose	250kg/m², heavy duty	300kg/m², masonry special duty
Duties	Inspection Painting Stone cleaning Light cleaning Birdcages and slung scaffolds also have this rating	Painting Stone cleaning Glazing Pointing Plastering	Brickwork Window fixing Mullion fixing Rendering Plastering Putlog scaffolds have this rating	Brickwork Blockwork Heavy cladding Putlog scaffolds have this rating	Masonry Blockwork Very heavy cladding
Maximum number of platforms	1 working platform	2 working platforms	2 working platforms + 1 at very light duty	2 working platforms + 1 at very light duty	1 working platform + 1 at very light duty
Widths using 225mm wide boards	3 boards (except birdcages and slung scaffolds)	4 boards	5 boards or 4 boards + 1 inside	5 boards or 5 boards + 1 inside or 4 boards + 1 inside	6 to 8 boards
Bay lengths	2.1, 2.4 and 2.7	2.0, 2.1 and 2.4	1.8, 2.0 and 2.1	1.5, 1.8 and 2.0	1.2, 1.5 and 1.8

Vertical spacings for platforms — putlog scaffold 1.35 m, independent up to 2.0 m

Design

All scaffolds must be planned. In principle all scaffolds must be structurally designed as well, but in practice standard approaches to layout are used by scaffolders without recourse to a specific structural design. The vast majority of scaffolds are erected in this way. Those which are formally planned on paper will normally be arranged by reference to the standard spacings currently in use for the duty to which they are to be put, but in some cases it is necessary to design specifically for a given set of circumstances. Then consideration must be given to the following loads

(a) the self weight of the scaffold structure
(b) the dead and live loads imposed on the scaffold structure during the course of the work; Table 1 gives examples
(c) loads from the wind, depending on the situation of the scaffold
(d) lateral loads.

Lateral loads may be positively identifiable but in all cases a horizontal force of 2.5% of the vertical load should be allowed for to cover the general inaccuracies of construction which will generate horizontal loads. Design of the scaffolding follows normal structural principles. The offset of each coupler reduces the capacity of the structure, but only by a few per cent. This is normally ignored, and the lower overall safety factor which results is considered adequate for a structure whose life is very short and whose construction can always be inspected. The new Code of Practice provides data for calculation, and gives some simple examples in the final section.[3] The best book available in English at the time of writing is Brand's *Falsework and Access Scaffolds in Tubular Steel*.[4]

Sheeted scaffolds require particular care in design, as the extra surface area creates much greater wind loads on the scaffold. This has been known to result in scaffolds being bent, and ties failing, as well as collapses.

Main risks

The user of the scaffold must not merely feel secure, he must be justified in this belief. His most frequent risks are of finding a guardrail missing or a gap in the decking. Incomplete scaffold should always be blocked off and labelled accordingly (Fig. 1).

Most scaffold failures are due to unsatisfactory tying. This may have been initially inadequate but more frequently it is caused by people removing ties temporarily and failing to reinstate them. Inadequate tying often occurs where a scaffold is sheeted to give weather protection. Another risk which has recently become more serious is direct overloading of scaffolds, as fork lift trucks as well as cranes now place brickpacks and the like on scaffolds, creating large local loads. These frequently overload traditionally dimensioned scaffolds, and specific areas of appropriate strength should be provided.

2. REQUIREMENTS

The legal position

In the UK the law requires that access of prescribed standards shall be provided. The Health and Safety at Work Act,

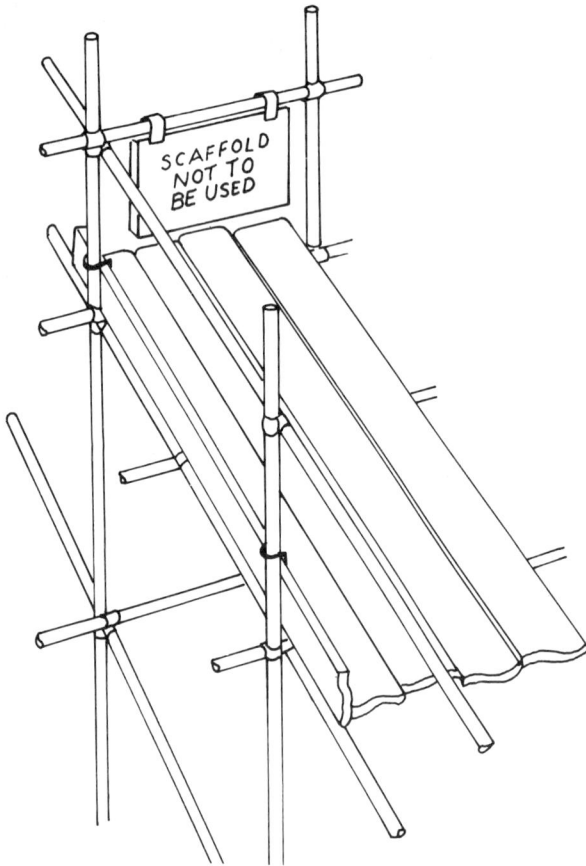

Fig. 1. Warn users of incomplete scaffold

1974 requires that there shall be a 'safe place of work'. This is quantified in the Construction Regulations[5-7] which have been in force considerably longer. They were written with reference to what was good practice at the time. The rest of this section is almost entirely controlled by these Regulations.

The basic need

Different operations create different loads and these different operations require different amounts of space for their satisfactory conduct. These vary from the requirement of one man to inspect some part of the construction, to the support of stonework for masonry. Where materials are stored temporarily on the scaffold platform, it is necessary to leave an adequate space for people to pass by to reach other parts. Table 1 gives information on this subject, and data on rule of thumb design for scaffolds up to 50m high.

The Regulations

The Regulations set down the legal minimum to be adopted in construction, although the possibility of varying some of the dimensions and requirements is made where it would otherwise be impractical to carry on the operation in hand. They set down the requirements in somewhat legalistic prose. A more readily assimilable presentation of them is contained in *Construction Safety*[8]. As well as the general philosophy that the working place should be safe, there are more specific requirements which interpret the meaning of 'safe'.

The regulations are all set out with imperial dimensions, and the figures given here are conversions. It is likely that rounded metric figures will be included when the regulations are next revised. The width of platform is incorporated in Table 1. There must be a guardrail at least 915mm (1m elsewhere in Europe) above the deck level of every platform from which it is possible to fall more than 1980mm, and there must also be a toe board rising at least 153mm above the platform. Brickguards, with openings small enough to stop a brick, can be used to protect the gap between guard rail and toe-board; in other countries an intermediate guard-rail is often mandatory. Timber scaffold boards of 38mm nominal thickness must not span more than 1524mm, and those of nominal 50mm thickness not more than 2590mm. However, several proprietary scaffold systems use thicker boards, as this has been found to give a more satisfactory rigidity.

The regulations regarding security on platforms apply when the possible fall is greater than 1.98 metres, but it is advisable to follow the rule throughout.

Platforms must be 'closely boarded'. This means that there should be no gaps. Chequer plating for example would be unacceptable, unless it is certain no one will be below. The trip hazard of overlapping boards is not allowed. Any alternative to timber boards should have a non slip surface.

Scaffold boards must not cantilever more than four times their thickness beyond their end support; but on the other hand they should not be so close to the support that a small movement could enable them to slip off it. If there has to be a ramp and if it is steeper than 1 in 4, then stepping battens have to be attached to provide a grip for people using it, although a gap of 101mm may be left for wheeling a barrow.

Ladders must continue 1067mm above the level which they serve, or there must be an alternative hand hold at this height for the operative reaching the top level. No ladder may continue longer than 9.144m without a platform on which a man may rest. Every ladder should be tied so that it cannot accidentally move (Fig. 2).

The regulations require that all scaffolds shall be inspected at least once a week and the record of the inspection be entered in the appropriate register.

This is in no way an exhaustive appraisal of the regulations. Designers and users must become fully conversant with them.

3. MATERIAL

General
Most scaffold is erected with materials which have already been used and are not new. Thus there is always a need to inspect them before erection to make sure that their condition is good enough. The following notes draw attention to some of the more significant aspects of quality which should be considered. See also the check lists. BS 1139[9] lays down the requirements for steel and aluminium scaffold components, and very little equipment is available which does not comply with it.

Steel tube
In the UK virtually all steel scaffold tube is 48mm in external diameter, 4mm thick, and made of grade 37 steel, lower than normal mild steel, grade 43. It may be galvanized, or have been supplied with the black transit coating from the mill, or have been subsequently painted. Tube should not be used if it is dented or creased, if it is visibly bent, or if it

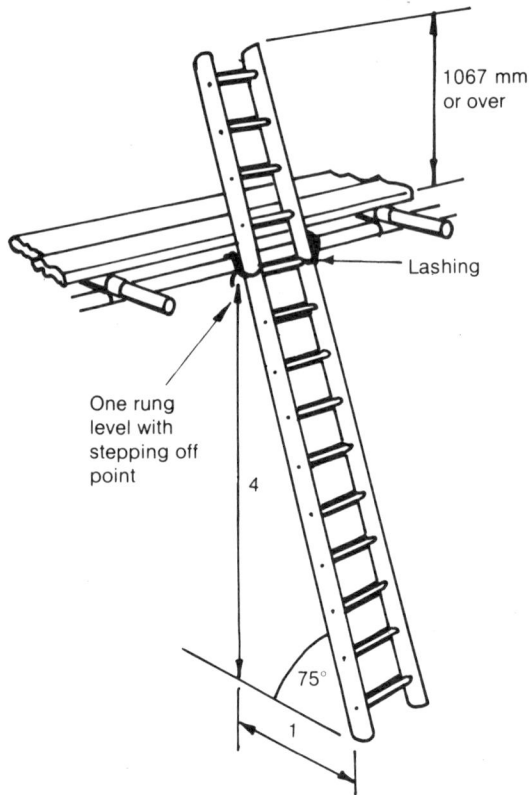

1067 mm
or over

Lashing

One rung
level with
stepping off
point

4

75°

1

Fig. 2. Positioning and securing of a ladder

is significantly pitted. If the ends show a noticeable reduction in thickness, a length 300mm long should be cut off and the remaining tube examined again. It frequently occurs that the end portion rusts internally, whereas the centre is little corroded.

Aluminium

Tubing made of aluminium 4.5mm thick is also used. It has the advantage of low weight in handling, and on occasion the reduction of load on the foundations makes it very advantageous. It is cleaner and more attractive. Because it has the same outside diameter as steel tubing it can be used with normal steel fittings, although it is possible to get aluminium ones for use, for example, where cleanliness is essential. Note that some prefabricated aluminium towers have tube which is 50mm outside diameter and special couplers are needed for these.

In use aluminium tube is nearly as strong as steel, but should not be mixed with it. This is because its deflection characterisitics are very different and a mixed structure could be unsafe. It is more expensive than steel, and considerably more vulnerable to theft.

Fittings

This word is used to describe all the components used with tube, although they can be sub-divided into couplers and accessories. Couplers positively attach one tube to another or to some other components. There are many manufacturers, some of whom provide equipment which complies with the British Standard, but with no margin, while others provide equipment which has a considerably better performance. Table 2 gives details of working loads.

Right angle couplers (Figure 3)

There are three types of coupler which connect two tubes at right angles, but the term right angle coupler is limited in its application to a coupler which has the strength requirements of the British Standard. This requires that the coupler shall have a failure load in excess of 1270 kg in slip. Normally a factor of safety of 2 is applied giving a safe working load of 635 kg. Different designs may look very dissimilar, but all perform the same function, and all provide at least the degree of security required by the British Standard. With the exception of the band and plate coupler, each has two bolts.

Putlog couplers (Figure 4)

A putlog coupler enables a short tube or board bearer to be fixed above a ledger. It does not project upwards, or it would prevent the boards laying flat. It is normally fixed by a single bolt. The required strength is small, but in an independent scaffold it is the coupler which connects the inner and outer lines of standards. Thus it is most important that it be in good condition.

Brace couplers (Figure 5)

A few putlog couplers designed many years ago provide an appreciably higher strength, approaching that of a right angle coupler. These couplers form a third group and some specialist firms may use them to attach bracing in less onerous situations. Check with the manufacturer if in doubt, as there is no easy means of identification.

Table 2. Safe working loads for scaffold fittings

Fittings which comply with BS 1139: (1964) will have the following capacities. It is necessary that they are in reasonable condition and properly fastened. They may be used on steel or aluminium tubes.

Right angle couplers		
Basic strength with slip prevented	15	kN
Slip along a tube	6.25	kN
Swivel couplers		
Basic strength with slip prevented	7.5	kN
Slip along a tube	6.25	kN
Putlog coupler		
Force to pull the tube axially out of the coupler	0.625	kN
Joint pins (expanding spigot couplers)		
Shear strength	21.0	kN
Sleeve couplers		
Bending strength (equal to the tube strength)	605 kN mm	
Tension	3.1	kN
Reveal pin		
Shear	0.625	kN
Load at the centre of a 1830mm long reveal tie	1.25	kN
Putlog ends		
Shear	1.12	kN
Adjustable base plate		
Axial load	30	kN

No other strengths are required by BS 1139, but fittings do have others.

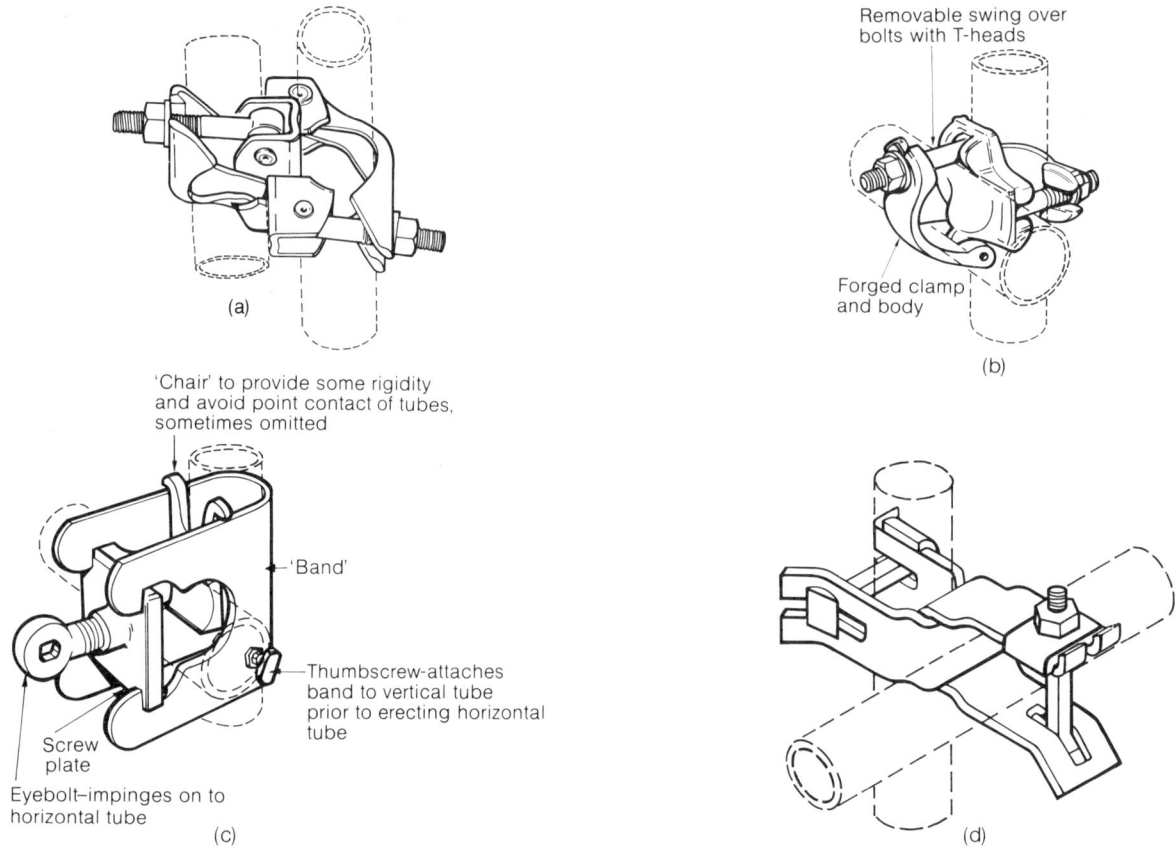

(a)

Removable swing over
bolts with T-heads

Forged clamp
and body

(b)

'Chair' to provide some rigidity
and avoid point contact of tubes,
sometimes omitted

'Band'

Thumbscrew-attaches
band to vertical tube
prior to erecting horizontal
tube

Screw
plate

Eyebolt–impinges on to
horizontal tube

(c)

(d)

Fig. 3. Typical right angle couplers; (a) pressed steel right angle coupler, (b) drop forged right angle coupler, (c) the original three piece universal coupler of 'band and plate', (d) spring steel right angle coupler

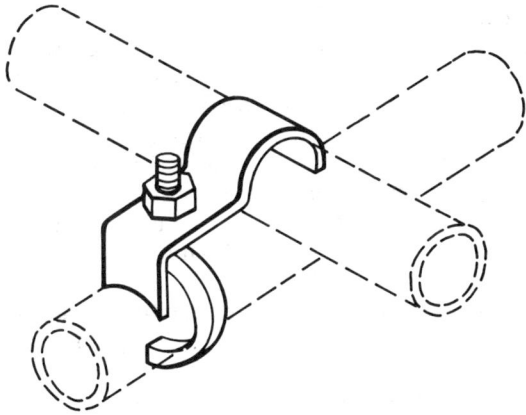

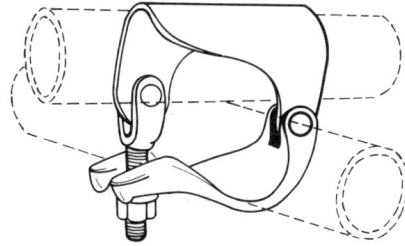

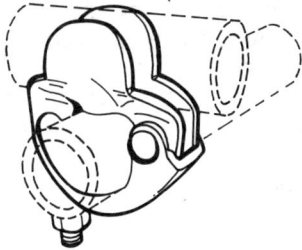

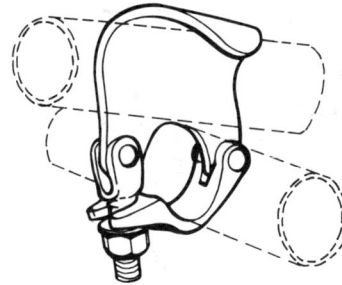

Fig. 4. Typical putlog couplers

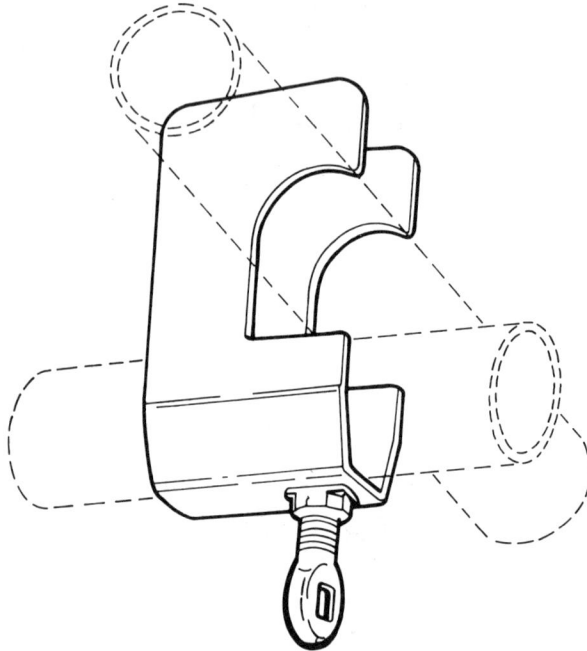

Fig. 5. Typical brace coupler

Swivel couplers (Figure 6)

Where it is impractical for two tubes to cross at right angles to one another, then a swivel coupler capable of rotating through 360° is very useful. It has the same slip capacity as a right angle coupler and the design is very similar, but it does not have the same rigidity as a right angle coupler, and thus should not be substituted for it except where this is unavoidable. Swivel couplers usually become unusable in the end because the rivet or swivel pin has become too slack or has come right out. Where there are significant loads it is essential that swivel couplers are carefully examined to ensure that the rivet or swivel pin is in good condition.

End-to-end couplers (Figure 7)

It is frequently necessary to have joints in tubes as the lengths available are insufficient. There are two main types of coupler. The spigot coupler goes inside the two tubes and is expanded to lock between them by turning the appropriate screw. Such a fitting cannot be expected to carry any load in tension, and the capacity in bending is very low. In an extreme case tightening can be so great that a poorly welded tube is split. The other type is a sleeve coupler in which the two tubes to be connected are gripped outside, and tightening of this fitting does not damage the tubes. It will permit a tension load of 3.20kgNf but its bending strength is not as good as uncut tube.

Base plates (Figure 8)

An ordinary baseplate consists of a 150mm square piece of steel plate with a small spigot in the centre to ensure that the vertical tube is located in the middle. In an adjustable base

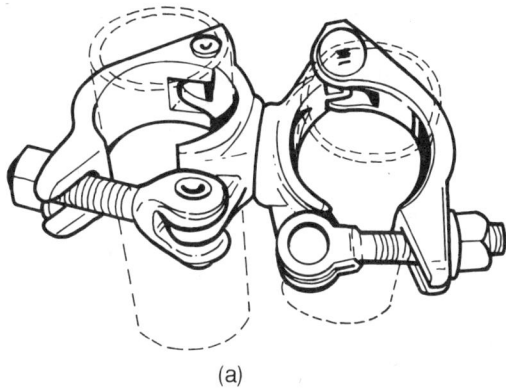

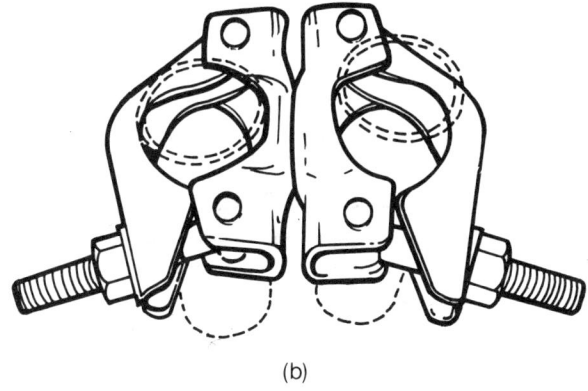

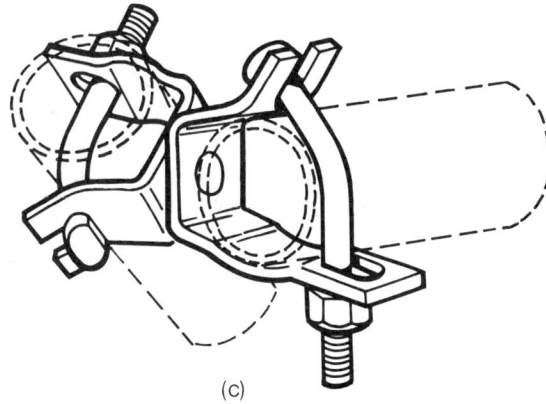

Fig. 6. Typical swivel couplers; (a) forged mild steel, (b) pressed steel, (c) spring steel

(a)

Band partially split to
permit freeing of one tube
while leaving coupler tight on other tube

Centring piece

(b)

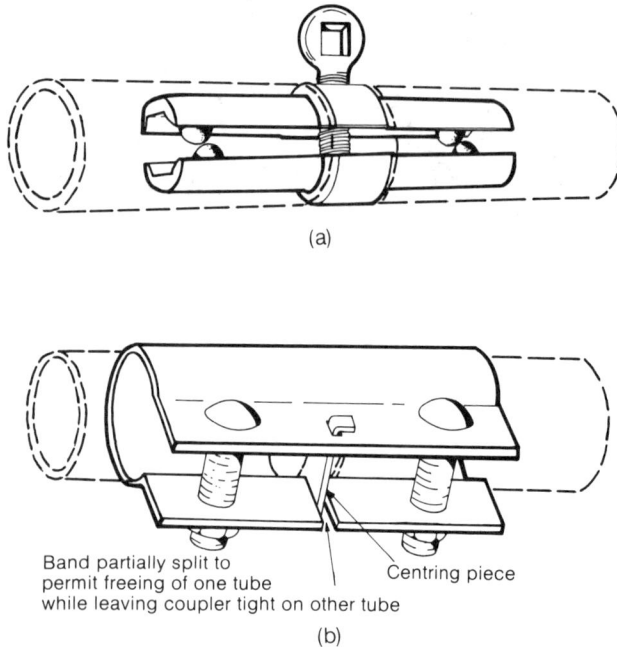

Fig. 7. End to end couplers; (a) typical joint pin (spigot coupler), (b) typical spring steel sleeve coupler

plate the central spigot is replaced by a screwed rod with a boss, over which the tube can be placed. Such a fitting is frequently essential with proprietary scaffold, and may also be used inverted at the top of the scaffold for load carrying purposes.

Reveal screws (Figure 9)

Where a tie cannot be put through the facade of a building, an alternative method of providing a fixing is to lock a tube between the two faces of a window opening or reveal. The small screw jack unit which goes into the end of the tube is called a reveal screw. It has an end which presses against the face of the opening and a threaded body with a boss which engages with the tube.

Other fittings

There are a variety of other specialized and less commonly encountered fittings but particular note may be made of parallel couplers enabling two tubes to be coupled side by side; joist couplers which enable timber or steel joists to be positioned at the top of the verticals; ladder clamps which enable the tops of ladders to be simply and effectively held; and toe board clips whose purpose is to fix the tops of toe boards.

Boards

The traditional British board is 225mm wide x 38mm thick. It is most frequently 3.9 m long, but other sizes are available. When timber is used for scaffold boards, protective metal ends are added to each plank. This serves to identify them clearly and reduces the likelihood of a board beginning to split. It also implies a warranty that the supplier has inspected the batch and removed those which are unsuitable.

British Standard 2482[10] provides for a board of a quality suitable for general use. However many boards are bought to other specifications, some of which are less onerous.

Boards should never be used if their condition is in doubt. They are frequently subjected to misuse, loading over too long a span for example, or left lying about on site so that mechanical transport can go over them. It is however extremely difficult to see any damage, and no convenient site test has yet been devised. Any board which is suspect should be put out of use, and care should be taken with all boards to see that they are not maltreated. Dynamic loading from

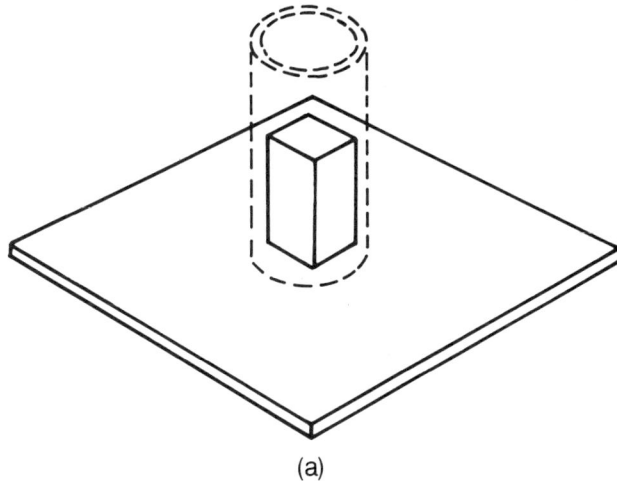

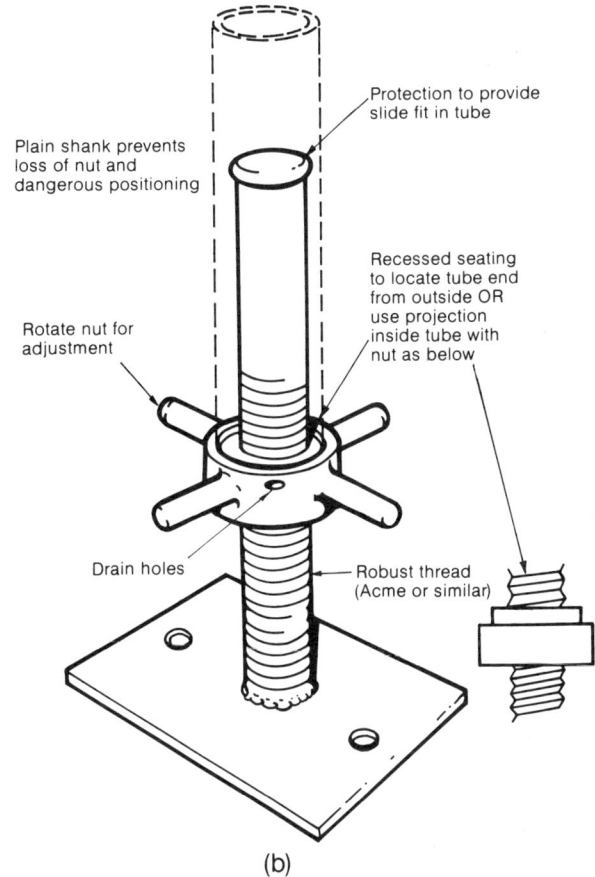

Fig. 8. Typical base plates; (a) plain base plate, (b) adjustable base plate

Points at corners prevent slip on packing

When using ensure nut does not approach too close to end of thread

Reveal pin

Timber packing

Tie should be attached to reveal tube at end opposite to the reveal pin and within 150 mm of the end

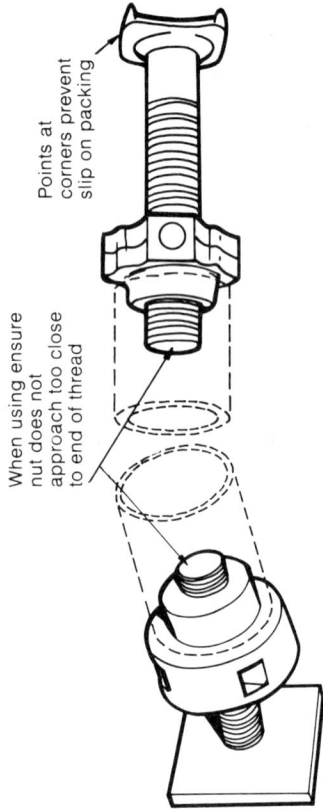

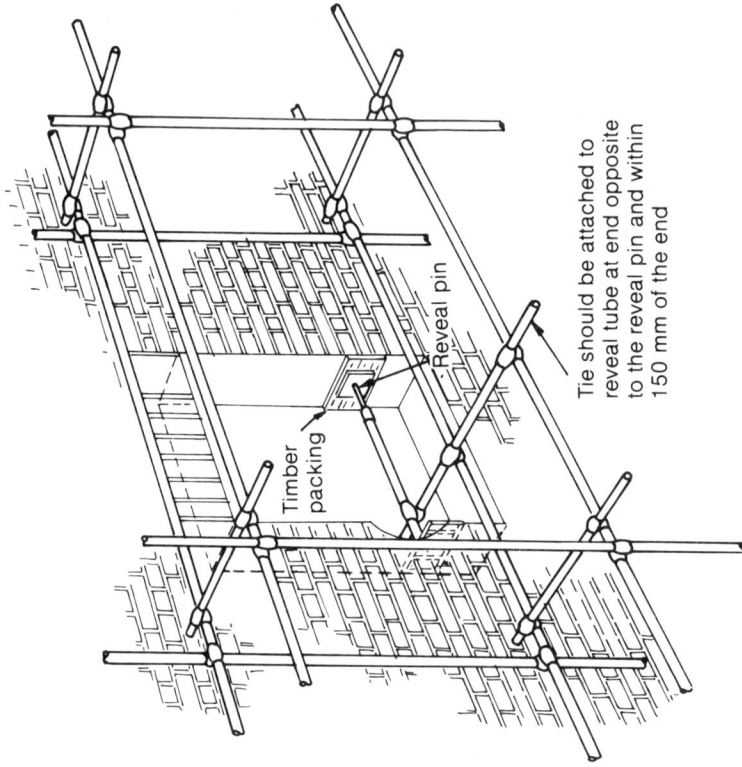

NOTE. The tube in the reveal can be in the vertical or horizontal position

Fig. 9. Reveal pins or screws and their use

14

people jumping down or material being thrown down can produce very high stresses.

Ladders

Most ladders in use are of timber, and BS 1129[11] is the relevant specification. Because of the possibility of flaws and damage being hidden, timber components may not be painted. Instead any protective coating must be transparent. Where damage has occurred ladders must be discarded or the unsuitable parts must be cut off.

4. FOUNDATIONS

No scaffold which has inadequate foundations will be safe. In most cases the loads coming down the standards are significant and must be carried by a surface which can safely withstand them. Settlement, particularly differential, must be avoided. Where a scaffold is to be set up on a concrete slab there should be little problem, though the strength of the slab must be checked. In some cases the scaffold will have to be set up on a pavement. Paving slabs are not particularly strong, and are often not very well bedded. Thus for the heavier scaffolds it is important that a substantial sole plate be placed under the base plates below the standards. Where pavements are sloping, some way must be devised to make sure that the bottom of the upright is correctly supported. This may be done by a small pad of mortar below the base plate to fill the triangular gap. A pavement may be considered level when it is within $3°$ of horizontal.

Where there is soft ground, only the lightest of scaffolds can be erected without removing the top soil, and it will thus frequently be necessary to do this down to the level of the sub-soil. Then sole plates such as sleepers may be placed to give a satisfactory support base to the scaffold. Thin boards give very little spread, as they are flexible. Where a significant spread is needed to apply the load to a wide area a board of minimum thickness of 75mm should be used. It is also important that they be bedded down to be in effective contact with the supporting surface without any gaps. Blaes, hoggin, or a very small amount of sand may be helpful, but for better results weak concrete is appropriate. The sole plate should be worked in to ensure good contact. If the sole plate has to bridge an opening, appropriate calculations should be done.

When ground is sloping, sole plates can be laid across the slope to give a level support. However there is a greater risk of the ground failing, and if it is steeper than 1 in 6 a careful analysis should be made, and a foundation design prepared. In all cases the ground should be examined for hazards. The most likely one is from badly consolidated backfill to trenches. There is also a risk that a trench will subsequently be dug too close to the foundation of the scaffold.

5. TIES

Theory

Ties should be placed sufficiently close together so that the scaffold structure is strong enough to span horizontally and vertically between them. Spacings which are normally

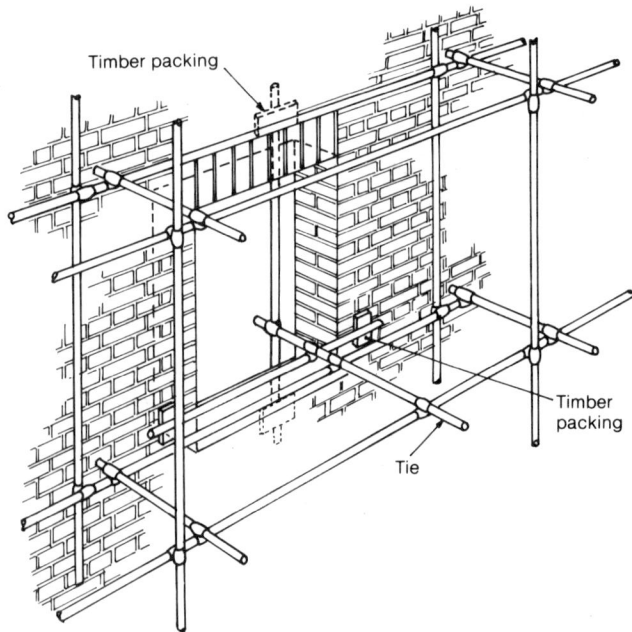

Fig. 10. *Through tie for independent tied scaffold*

Labels in Fig. 10:
- Timber packing
- Timber packing
- Tie

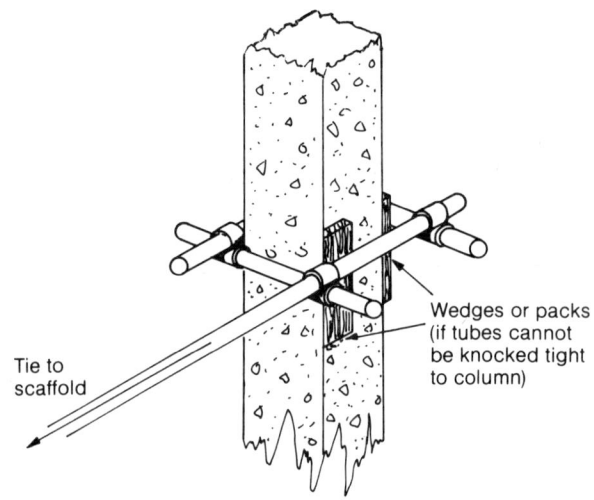

Fig. 11. *Box tie*

Labels in Fig. 11:
- Tie to scaffold
- Wedges or packs (if tubes cannot be knocked tight to column)

16

satisfactory are given in the next two sections. The normal capacity inwards and outwards is 635kgf, given by a single coupler. While the basic capacity of a tie is normal to the surface served, inevitably some capacity vertically and some capacity laterally along the building will also be provided. It is normal to arrange the details of the scaffold so that this capacity is not called on unless there is a catastrophic failure of the scaffold. Any force tangential to the facade should be catered for separately.

It is essential that an adequate number of ties is present at all times. As the scaffold is constructed, it must be tied in, and if a tie has to be moved to enable some construction operation to proceed, an alternative should be provided before the one to be taken out is removed. Where a new building is being constructed, thought should be given at an early stage to the provision of anchorages, as these can often be provided during the construction much more satisfactorily than is possible with an existing building.

Types
It is necessary to prevent the scaffold falling away from the structure, and likewise to prevent it moving inwards. Thus it is usual to have a rigid tie member which can be positively connected to the structure, usually a piece of scaffold tube. There are various ways of attaching the scaffold.

Reveal ties (Figure 9)
When scaffolding existing structures it may be impractical to go through windows. An alternative anchorage is to use a short length of tube and a reveal screw which is tightened between the sides of a window opening. The scaffold is tied

to this with another tube. It is normal to place timber packers at each side, to reduce the risk of damaging the window opening, but this may increase the risk of the tie failing. It should only be used when other ties are impossible, as its strength is low. Consideration should be given to providing fixings for subsequent maintenance.

Through ties (Figure 10)
In this case the tie goes horizontally through a window or other opening. Tubes are placed at the outside and the inside of the opening and are fixed with right angle couplers.

Box ties (Figure 11)
In this case two ties go either side of a column and two lateral tubes prevent inward or outward movement.

Lip ties (Figure 12)
This type is less satisfactory, because it does not tie fully round the structure. It consists of a tie-tube and two short lateral tubes one in front and one behind the facade. They should be used in handed pairs, to prevent movement in both directions.

Drilled anchorages (Figure 13)
It is possible to drill and fix a screwed eye bolt into the facade. If the facade is of concrete, ample strength will almost always be obtained. If of brickwork, it will be necessary to check that the expected capacity is available. The preferred detail is to use an eye bolt whose internal diameter will just accommodate a short scaffold tube to which the tie is fixed, but if a smaller eye bolt is used a com-

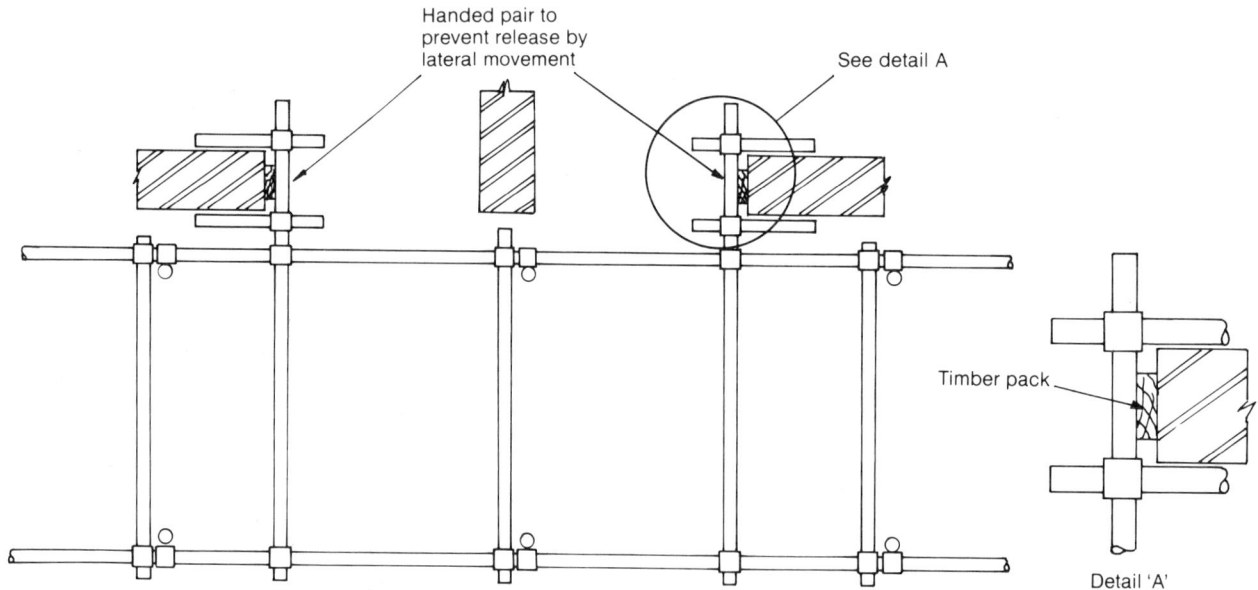

Handed pair to
prevent release by
lateral movement

See detail A

Timber pack

Detail 'A'

Fig. 12. Lip ties

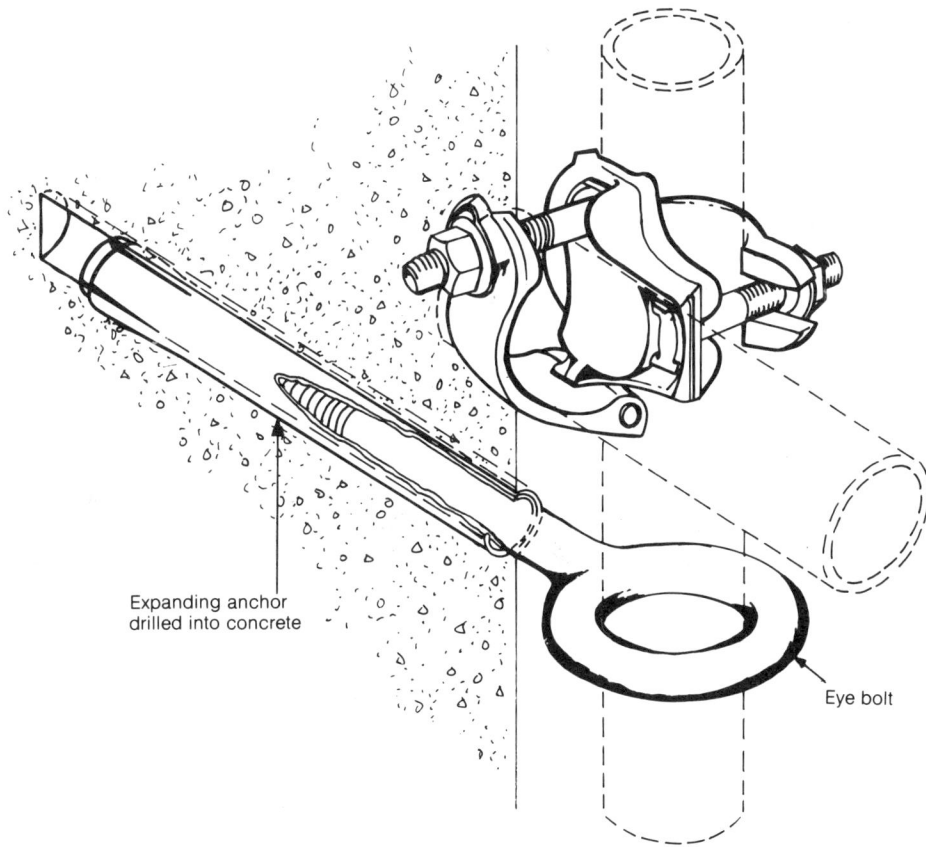

Expanding anchor
drilled into concrete

Eye bolt

Fig. 13. Drilled anchorage with eye bolt

bination of a wire band and a tube acting as a strut only can be used. Alternatively a short length of tube with a washer and long bolt may be appropriate.

Cast in anchorages (Figure 14)
A similar arrangement can be achieved when a building is being built, by arranging to cast in a suitable pattern of threaded sockets. For a brick clad building, it should be possible to lay these out so that only one brick need be left out per tie, to be completed when the scaffold is dismantled.

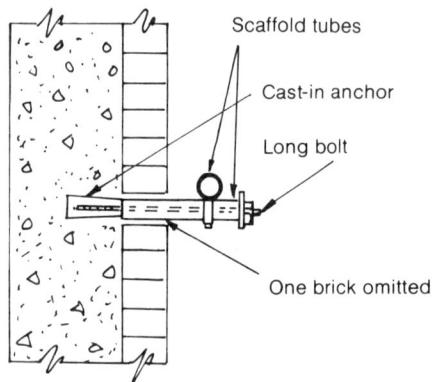

Fig. 14. Cast-in anchorage

6. INDEPENDENT SCAFFOLDS

Description (Figure 15 and Table 1)
Despite its title, an independent scaffold requires ties to give it lateral stability. However no vertical load is transferred to the building. It comprises pairs of standards typically 1.25 m apart and spaced along the building between 1.2 and 2.4 metres apart. The pairs of standards are connected horizontally parallel to the building with horizontal tubes called ledgers. These are normally spaced vertically at the working height of 2 m or just under. The inside and outside grids so formed are connected with short tubes known as transoms. Ideally these should be placed adjacent to the standards but connected to the ledgers. However for reasons connected with board lengths, it is accepted practice to position them up to 300mm from the standards, if this will save duplicating horizontal tubes. Because 38mm boards should not span more than 1.5 m, additional short tubes known as board bearers are put between the transoms to provide suitable support for the boards and their ends. A layer of ledgers, transoms and board bearers is referred to as a lift. The scaffold is completed with ledger bracing (a diagonal tube joining the front ledger at one level with the back one above or below) and facade bracing. This latter runs diagonally up the facade to provide stability along the structure.

Details
Standards may be lengthened using spigot couplers unless there is a risk of uplift, but such joints should be staggered so that they do not occur at the same level in adjacent

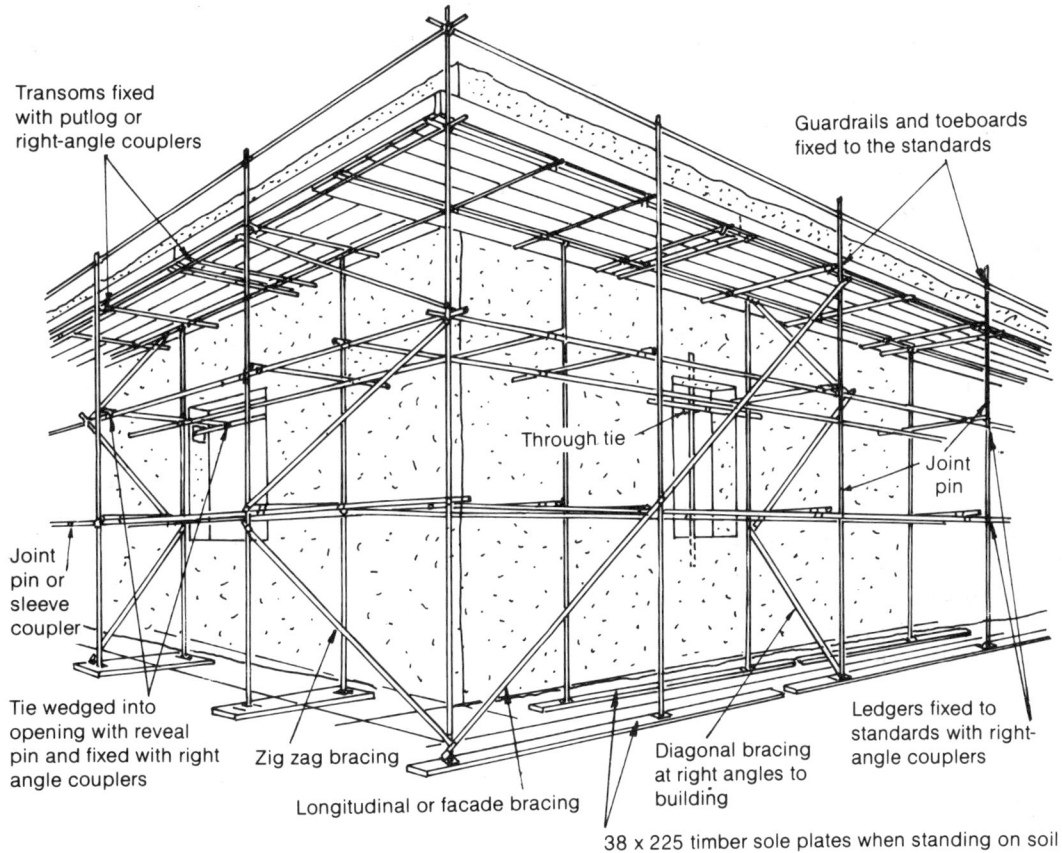

Transoms fixed with putlog or right-angle couplers

Guardrails and toeboards fixed to the standards

Through tie

Joint pin

Joint pin or sleeve coupler

Tie wedged into opening with reveal pin and fixed with right angle couplers

Zig zag bracing

Longitudinal or facade bracing

Diagonal bracing at right angles to building

Ledgers fixed to standards with right-angle couplers

38 x 225 timber sole plates when standing on soil

Fig. 15. Independent tied scaffold

standards. As a tube with a sleeve coupler is weaker than an uncut tube, the joint should not be halfway between node points. Ledgers should be fixed to the standards using right angled couplers. Joints in ledgers should preferably be made with sleeve couplers, and the same comment about position applies. The transoms are normally fixed using putlog couplers, as are the board bearers. It is normally not practical to use right angle couplers for these purposes, as they cause the deck to be uneven. Where a level of boards is removed, intermediate board bearers may be removed, but not transoms. However it is better and often cheaper to let them all remain as normally the deck will have to be replaced when the final part of the work is in progress. If there is a risk of the wind blowing the boards off the scaffold, they should be fixed down.

Bracing

The purpose of all bracing is to provide rigidity and strength. The classic principle to follow is that of triangulation. Each diagonal member should ideally connect to two node points in the structure. Because of the offset inherent in the construction of tubular scaffold, this is not actually possible but every effort should be made to attach members as near as possible to the ideal point. Where a diagonal is put across a rectangle but away from the node point, instead of all the members being in simple compression or tension, bending is introduced.

The ledger bracing should be connected from one ledger to the one above on the other side with right angle couplers. If this is arranged from the lower outside to the upper inside ledger, it will conflict with the toe board. A solution to this problem, sometimes adopted, is to use a swivel coupler for the lower connection and to put it above the toeboard, but this is undesirable. Ledger bracing is normally put on alternate pairs of standards.

Facade bracing starts at the bottom and may either continue in one long diagonal line to the top, or alternatively it may rise past only one or two lifts. It then starts again with the original standard and goes across as often as is necessary to reach the top. Ideally such bracing should be connected to projecting transoms very close to standards. Where this is not possible, swivel couplers may be used, but particular attention should be paid to putting them as near the node points as possible.

Braces should be connected to projecting transoms with right angle couplers. This is only practical if the scaffold has been planned with this in view. The layout of transoms is largely dictated by the length and disposition of boards, but transoms to be braced must form a diagonal line adjacent to the various standards to facilitate the bracing. At the bottom the brace should continue below the lowest transom to ground on the base plate, otherwise the lowest part of the standard has to carry all the horizontal forces. Joints in braces should be made with sleeve couplers, because they may be in tension or compression.

Ties

Such a scaffold is not stable on its own when more than 3 m high. Ties should be provided at the rate of 1 for every 33 square metres of facade, where it is certain they will not require any adjustment during the life of the scaffold. However, should this be anticipated, a rate of 1 per 25 square metres should be adopted. A replacement tie or ties should always be put in before one is removed.

Access at the base
It is often necessary to erect a scaffold on a pavement or some other place where access through is desired. Where the lower part of the scaffold is unbraced, its capacity is reduced. Thus where it is desired to have access through such a scaffold, the most satisfactory engineering solution is to board out the lowest lift 100 or 200mm from the bottom, with ramps at both ends. Where this is impossible, a check should be made on the reduced capacity of the standards. Where a scaffold stands on unlevel ground, the lowest level of ledgers and transoms should not be more than one third of the basic lift to lift height from the ground.

Vertical access
Attempts to provide access by stairs to higher working levels do not attract the success they would appear to deserve, and so the basic method remains the ladder. It should have both stiles (the side members) set firmly on the ground. It should lean at about 1 to 4 and it should be tied near the top to prevent it moving. Other details are given under 'The Regulations' (page 4 and Fig. 2). Sometimes a ladder is placed leaning against a scaffold, but then it is necessary to get past the guard-rail — by ducking or climbing. It is preferable to build an extension of the main deck as a separate ladder tower with the ladder coming through the floors. Openings should be kept as small as practicable. When it is impossible to secure a ladder a second man should remain below to prevent it from moving.

7. PUTLOG SCAFFOLD

Description (Figure 16 and Table 1)
This type of scaffolding has a long history, although its use is diminishing. The support comes half from a line of standards and half from the building itself and so it is not independent. A single line of standards, spaced from 1.5 to 2.1 m apart, is erected about 1.2 m from the structure. Ledgers are attached to these standards, with a lift to lift height of about 1.35 m.

Near the standards are attached putlog tubes. These are lengths of tube about 1.5 m long, with one end flattened. This flattened end is placed on the brickwork that is being constructed, or very occasionally into existing brickwork. Other intermediate putlogs are used as board bearers. Facade bracing should be provided. It frequently happens that there are windows, or some similar difficulties, which interfere with the support of the ends of the putlogs, and alternatives have to be adopted to support the inner end from a bridle tube supported by the adjacent putlogs to right and to left.

Construction
The ledgers are connected to the standards using right angled couplers. The putlogs are connected to the ledgers using putlog couplers. The facade bracing should preferably be connected to the outer ends of the putlogs using right angle couplers, but alternatively swivel couplers to the standards may be used. In either case the bracing should go as close to the intersection of standard and ledger as possible.

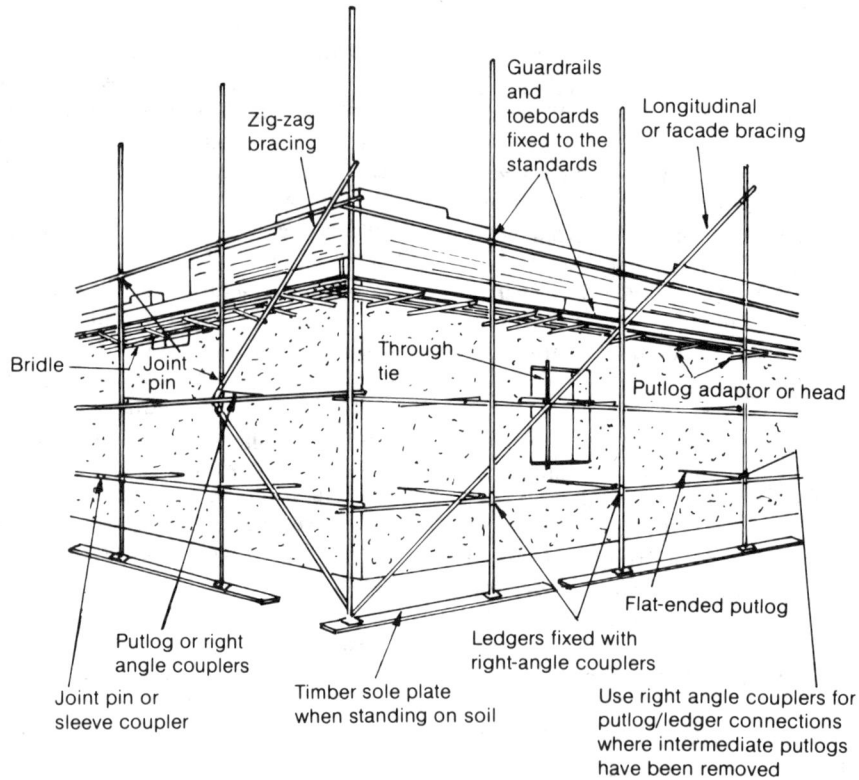

Zig-zag bracing

Guardrails and toeboards fixed to the standards

Longitudinal or facade bracing

Bridle

Joint pin

Through tie

Putlog adaptor or head

Putlog or right angle couplers

Flat-ended putlog

Joint pin or sleeve coupler

Timber sole plate when standing on soil

Ledgers fixed with right-angle couplers

Use right angle couplers for putlog/ledger connections where intermediate putlogs have been removed

Fig. 16. Putlog scaffolds

Tying

As the inner support to the scaffold is the structure, and any outward movement of the scaffold would result in its failure, ties are even more important. They should be provided at a rate of 1 for 25 square metres. It is also necessary that at least 50% of these be positive ties. Positive ties are those which have a resistance to movement which does not depend on friction.

Birdcage scaffold (Figure 17)

Where access over an area is required, for example for plastering a ceiling, a grid of standards is used to support a deck. The standards will normally protrude through, preventing the boards from meeting along that line. Where this gap is a particular problem, hardboard may be used to close it.

Suspended scaffold (Figure 18)

It is useful to have a working surface which can be adjusted vertically to the precise height required. This is achieved by outriggers, fixed higher up the building, from which platforms are suspended by ropes, and which are usually capable of being raised and lowered. These are almost invariably provided by specialist subcontractors. It is vital that the entire load carrying path has a large factor of safety, from the deck right back to the outrigger fixing, and the structure itself. There is no alternative load path, should something fail. The British Standard code of practice for Temporary Suspended Scaffold[12] gives information. Also available is a Users' Guide to Temporary Suspended Access[13].

8. MOBILE TOWERS

Description (Figure 19)

Scaffold can be dismantled, carried to a new location and re-erected, but where it can be moved on wheels, this may be a considerable advantage in time and cost. The industry uses mobile towers constructed either from purpose made components or from ordinary scaffold parts. Each tower has a

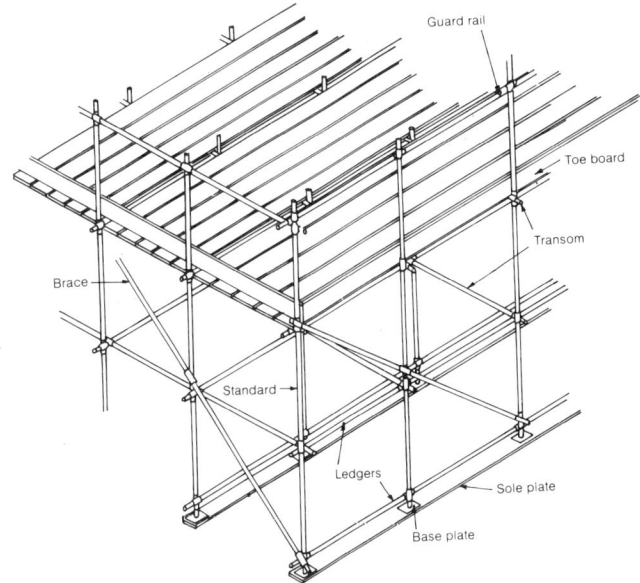

Fig. 17. Birdcage scaffold

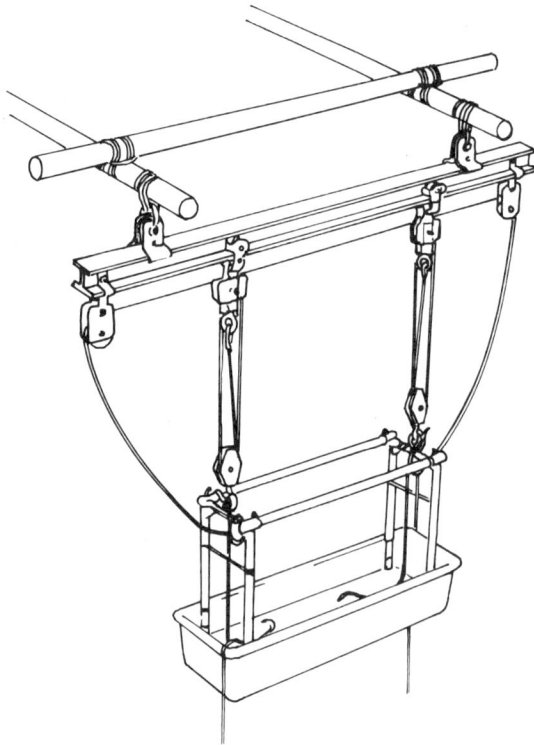

Fig. 18. Suspended scaffold

platform at the top and possibly at other levels. It preferably has inbuilt access in the shape of a ladder or stairways. In some cases it will have outriggers to widen the base to permit a greater height with adequate stability. Such scaffolds are used for dealing with services at ceiling level, for painting, for various other operations to facades, and frequently for stripping soffit formwork.

Construction

A mobile tower is constructed on four or more castor wheels. The wheels must be capable of being locked to prevent turning, and preferably the castors should also be capable of being prevented from rotating. The body of the scaffold will normally be a plain rectangle with decks at one or more levels, with guard rails and toe boards as for a normal static scaffold. The limiting proportion for a mobile scaffold up to a height of 4.5 m is 3½ to 1, or if it is made of aluminium, 3 to 1. This is a rule of thumb to prevent overturning. For higher scaffolds individual calculations should be made, and in general it is necessary to tie them to the structure to provide the necessary stability. Horizontal forces arise from various operations, and small towers are a particular risk.

Use

Before climbing to the top of a tower, the operative should ensure that the brakes have been applied. It is essential that no attempt is made to move the tower from the top, or while anyone is on it, and it is desirable that there is minimal loading on the platform when it is being pushed from the bottom. Care should be taken that it sits on an even, level surface. (Adjustable castors are available, which will enable a tower to be plumbed on a surface slightly out of level). Towers, particularly those of lighter construction, have in practice a fairly limited capacity and are vulnerable to overturning. Reference should be made to the manufacturer, or to the rules in the Operators' Code of Practice.[14]

Proprietary scaffold towers
In all cases the manufacturer's rules must be followed.

Access
Some proprietary towers are manufactured with horizontal members which can be climbed like a ladder. However, this is generally undesirable, as it almost invariably means climbing outside the tower. Almost all proprietary towers provide trap doors at the top and other levels, and a ladder system within the tower. This reduces to a minimum the risks of overturning or of people falling.

9. PROPRIETARY EQUIPMENT

The next development after the success of tube and fittings was the use of frames. Each comprises two standards, 1.5 or 2.0 m high with a transom, all as one prefabricated unit. They are assembled one above the other and the horizontals are formed of tube. Outside the UK many such systems have been developed further by replacing the horizontal ledger tubes with components purpose designed to give the scaffold longitudinal strength and appropriate guard-rail facilities. This approach is not very popular in the UK. Instead most proprietary systems are based on single components with

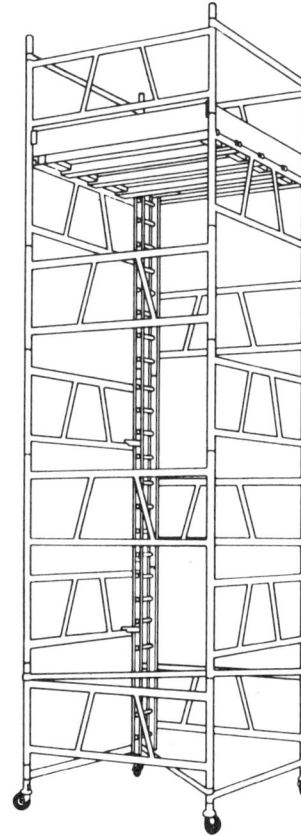

Fig. 19. Typical mobile tower

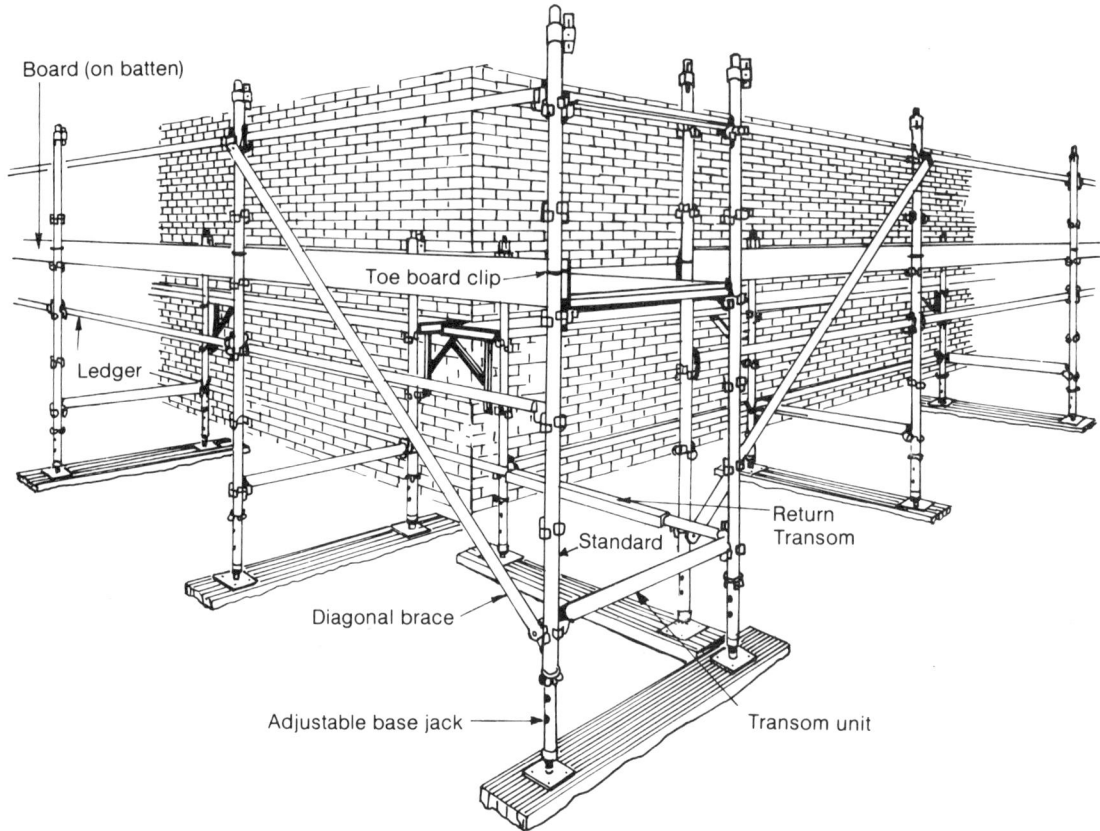

Fig. 20. Typical proprietary scaffold

inbuilt couplers conceived for easy assembly. These are lighter and less bulky for storage and transport (Fig. 20).

Standards, typically 2.0 m high, have at half metre intervals a series of attachment points. To these are fixed transom members of appropriate length to give the correct width for the scaffold, and comparable ledger members. In some systems additional board bearers enable standard boards to be used. In others a thicker board is used spanning, and located between, the transoms at the standards. Some systems, particularly those which have been developed over many years, have many special parts, giving them considerable versatility. Despite the similarity of concept, it is the exception to be able to use components of one system with another.

A proprietary scaffold is closer to the ideal structure, as most members are in line. In most cases the joints themselves are stiffer, and can carry greater moments. Thus the amount of bracing needed is less, and access along the scaffold is less impeded. In practice the choice between proprietary and tube and fittings scaffold is usually commercial. Unless the building to be serviced is very awkward, either type can be used. If there has to be adjustment to platform heights and so on, it is frequently easier to use proprietary scaffold. But the cost aspect is the most important. Due to lighter components and simple joints, proprietary scaffolds can be put up with less labour than tube and fittings, although the capital cost is greater. Thus for scaffolds which will only be up for a short time, the proprietary scaffold will prove more economical, while the scaffold which has to stand for many months will be cheaper made in tube and fittings. One further factor is the particular skills available. Provided the layout is straight forward, the level of ability needed to erect proprietary scaffold is less than for tubular scaffold.

While the principles in this guide apply equally to proprietary scaffold, the rules provided by the manufacturers of the particular scaffold must be followed. Apart from complete systems, manufacturers offer components for use in conjunction with tube and fittings. Examples are fabricated trusses, and transoms with two or four half couplers welded on. As in the case of complete systems, the manufacturers rules must be followed.

10. QUALITY

General
Scaffolding is an engineering structure. In a well built scaffold, the intersection of tubes at a node point creates a small eccentricity. Investigation into the effects of this indicates that it is of little importance, so long as the eccentricity is limited to about 75mm. But it is important that it is not permitted to be significantly greater than this figure. Thus all fittings which connect to a common tube should be placed as close to each other as is possible. Similarly, as the concept of bracing is to triangulate, the accuracy of positioning is fundamental.

Dimensions
It is clearly important that the dimensions adopted in the design are followed closely, and that the correct numbers of tubes and other components are used. Individual tubes should be horizontal or vertical to an accuracy of 25mm in 6 m, and never exceed 100mm from their correct position.

Quality of components

Even new components are sometimes faulty, and the following points should be examined.

Tube

This should be straight to the eye and have no dents or signs of creasing where straightening has taken place. It should be free from rust pitting, and the ends examined to see that they are still of the appropriate thickness.

Couplers

All couplers should have their working parts operating satisfactorily. Nuts should move along threads using the fingers only. Swivel couplers should have the rivet in good condition with a minimum of movement between the two parts.

Equipment should never be thrown down, as this is dangerous as well as causing damage to the material.

11.   MAINTENANCE

This is best done off site but the following notes may help those on lengthy jobs.

Tubes can be straightened, either by a jim crow or preferably by reeling using a tube straightening machine. Any part of a tube which is dented, creased or split should be cut out. Tube straightening machines usually incorporate cleaning systems as well. 'Black' tube requires paint to protect it, as also does galvanized tube when the zinc coating gets damaged.

Fittings should be cleaned, and threads checked. Timber boards should be cleaned and carefully examined for incipient cracking. They should not be 'tested' by jumping on them, as this can cause damage. Similar remarks apply to timber ladders.

Minor straightening of steel or aluminium boards or platforms can be undertaken at site, but this is best dealt with at the depot.

Proprietary equipment may likewise be given minor straightening. It is important to look at all welds carefully, and if anything needs doing, send them away.

12.   CHECKING

Before use a scaffold must be checked to ensure

(a)   that it is in accordance with the design
(b)   that it is satisfactory for the purpose for which it has been erected
(c)   that it has been put together properly
(d)   that a responsible superviser has signed the register.

Point (c) covers the accurate positioning of the components, and the satisfactory tightening of all bolts, and so on. In some cases, comprehensive drawings will exist, and can be used as the basis of the check. However, frequently the information will be much more informal, in some cases only verbal.

The first point to establish is whether the use to which the scaffold will be put will be catered for adequately by the design. Points to be considered are

(a) appropriate levels of decks from which to work

(b) adequate access, both horizontally and vertically

(c) freedom from 'traps' such as awkwardly placed access openings

(d) positioning of edges of working decks neither too close to nor too far from the actual work

(e) the actual loads should not exceed those used in the design.

Until experience is gained it is necessary to check a large proportion of any scaffold. Later a spot check will reveal the general state of erection, and further checking can be carried through should this be indicated by the first sample.

Be aware that it is too late in the day to find that base plates are bent when the scaffold is up three lifts. For this and similar reasons it is essential that checking is a continuing process during erection to reduce the waste of effort should faults be found. Constant checking also reduces the difficulties should a component be in a borderline condition for use, as it is simple to replace it at the point at which it is being initially erected, but once it is incorporated in the scaffold it becomes very difficult to do this.

Points to consider include the adequacy of access and problems where scaffold erection is still proceeding for later work, where it is essential that a barrier be created to prevent people inadvertantly going beyond the completed section. The initial and the subsequent weekly inspections are frequently undertaken by the senior scaffolder on site. It is usually he who will sign the statutory register and enter the record of the state of the scaffold. But this duty may be undertaken by an engineer.

Interim checking

It is fairly simple to modify a scaffold, and whilst this is the exclusive duty of a scaffolder, it is not infrequent that others may make some change to suit their immediate convenience, or borrow a component to use somewhere else. Thus it is necessary that weekly inspections are undertaken to ensure that the scaffold is still in proper condition. Points to look for are the continuing presence of guard rails, toe boards and the full complement of ties to the building. A weekly entry in the scaffold register is required.

Detailed information for checking is given below.

Checking the materials

See also notes under the descriptions.

Steel tubes

To be straight. If in doubt, support the two ends on horizontal timbers, and roll the tube.

To be free from dents and creases. Minor dents from the bolts of band and plate couplers are acceptable. To have full material thickness at the ends, without a split.

To be free from excessive corrosion and pitting. The ends should be cut square, cleanly.

Fittings

All should be free from excessive or loose rust. Each nut should be moveable along the whole length of its thread with the fingers. Any distorted fittings should be discarded. The inside of a coupler should conform to a tube. Nut flats should be undamaged.

Swivel couplers

The rivets tend to loosen with time. Slack couplers should be discarded. They should be able to swivel through 360°.

Scaffold boards

To be free from concrete, etc. so that an inspection can be made. There should be virtually no warping or bowing. Splits from the ends not exceeding 300mm are acceptable. The ends must be protected with metal strapping correctly fixed. Any boards which are cracked or appear to have been damaged in any way should be discarded, and treated to prevent their re-use.

Timber ladders

To be free from cracks, damaged or loose rungs; to be straight, clean, and particularly if shortened, the two stiles to be the same length. When put on a flat surface, wind or twist should not exceed 10mm.

The basic layout

The dimensions should tally with those in the design to ±25mm, so that the scaffold is in the intended position in plan and elevation.

The correct quantity of material shall have been used. This is particularly important where detailed designs have been done.

In the case of outline designs, is there enough bracing and tying in?

Standards and ledgers should be true to line within 1 in 100.

Are there any special features — fans, local sheeting — which could cause loads which have not been anticipated?

Foundations

Are the base plates level?

Unless on concrete, is there a suitable sole plate? Are the base plates centralized on the sole plate? Is the sole plate properly bedded?

Is the ground below strong enough? (Or is the structure, if it is the support?)

Is there a risk of differential settlement?

Are there weak spots such as manhole covers or recent backfilling of trenches?

Are all essential services like hydrants and beer cellar entrances kept clear?

Tying

Are there adequate provided?

Will each tie safely carry 635kg?

Will the structure safely support 635kg loads imposed by the tie?

Are the ties adequately connected to the scaffold proper?

Are there adequate through ties in putlog scaffolds?

Details

Are connections correctly made? Couplers at node points should touch each other as far as possible.

Main connections should be made with right angle couplers. Bracing should be fixed with right angle couplers, wherever possible.

Are end to end joints properly located, and made with the appropriate coupler?

Are all toe boards and guard-rails properly fixed? Are boards supported sufficiently? In particular is there a support within 150mm of each end? Are short boards prevented from being upended by someone standing on the end? Are ladders properly footed, tied at the top and is there a hand hold above the top, where the ladder does not rise 1067mm above the deck?

Where a torque wrench is available, check a number of nuts. Once confidence in a gang is established a smaller sample will be sufficient. Where a torque wrench is not available, use the manufacturers spanner, but pull on it so that you could undo the nut again, not with all possible force! A typical torque is 40Nm but manufacturers' data should be consulted.

Are putlogs properly seated in the brickwork?

Are joints adequately staggered?

Periodic inspections

Have modifications been made? If so are they satisfactory? In particular are the ties still in proper order? Is all the decking complete, and are all guard rails and toe boards in position?

Have materials or equipment been placed to overload the scaffold?

Are all ladders still properly positioned?

Have the foundations deteriorated, or been put at risk? Has the scaffold deflected unduly or moved at any point? Has any damage been caused by cranes or other plant and not been put right?

Proprietary scaffold

All the points about ordinary tubular scaffold apply to proprietary scaffolds too. The material in some cases is more vulnerable than tube and fitting, so a careful look is even more important. Welds are used in fabrications, and these can crack. Minor bends in the joint components may result in difficulties in assembly, cured with a heavy hammer, which in time does further damage.

In some cases imperial and metric components are of very similar dimension, and care is needed to ensure the correct one is used.

ACKNOWLEDGEMENTS

The author wishes to thank the following sources for their kind permission to use the relevant illustrations in the text.

National Federation of Building Trades Employers, 82, New Cavendish St., London, WIM 8AD. Figures 1 and 18 were reproduced from *Construction Safety*, published by BAS Management Services for the NFBTA.

McGraw-Hill Book Co. (UK) Ltd. Figures 3(a)-(c), 4, 5, 6, 7, 8, 9, and 16 were reproduced from *Falsework and access scaffolds in tubular steel* by Brand, 1975.

Stephens and Carter, Ltd., Turriff Building, Great Western Road, Brentford, Middlesex. Figure 19 was reproduced from the *S & C guide to access scaffolds*.

British Standards Institution, 2, Park St., London, WIA 2BS. Figures 9, 10 and 15, were reproduced from CP97 Part 2, 1967. Complete copies may be obtained from the BSI. Tables 1 and 2 are based on the new draft code on scaffolding, 80/11872.

Kluwer Publishing Ltd., 1 Harlequin Avenue, Great Western Road, Brentford, Middlesex, TW8 9EW. Figure 17 was reproduced from the OCPA *Safety Manual*, © Oil and Chemical Plant Constructor's Association, by courtesy of Kluwer Publishing Ltd.

C. Evans and Sons Ltd., Evans House, 78-82 High St., Brentwood, Essex. Figure 20 was reproduced from *Universal scaffolding and propping systems.*

REFERENCES

1. BRITISH STANDARD CODE OF PRACTICE. *The use of safety nets on constructional works.* CP93, BSI, London, 1972.
2. DRAFT FOR DEVELOPMENT. *Performance requirements for access and working scaffolds.* DD72, BSI, London. In preparation.
3. BRITISH STANDARD CODE OF PRACTICE. *Metal scaffolding. Part 1: Common scaffolds in steel. Part 3: Special scaffold structures in steel.* CP97, BSI, London, 1972. New edition begin drafted as BS5973.
4. BRAND R.E. *Falsework and access scaffolds in tubular steel.* McGraw-Hill, Maidenhead, 1975.
5. HMSO. *The construction (general provisions) regulations, 1961.* No. 1580, HMSO, London, 1961.
6. HMSO. *The construction (lifting operations) regulations, 1961.* No. 1581, HMSO, London, 1961.
7. HMSO. *The construction (working places) regulations, 1966.* No. 94, HMSO, London, 1966.
8. BUILDING ADVISORY SERVICE. *Construction safety.* BAS, London, 1978.
9. BRITISH STANDARD DOCUMENT. *Specification for metal scaffolding.* BS1139, BSI, London, 1964.
10. BRITISH STANDARD DOCUMENT. *Timber scaffold boards (38mm x 225mm softwood).* BS2482, BSI, London, 1970. Under revision.
11. BRITISH STANDARD DOCUMENT. *Timber ladders, steps, trestles and lightweight staging for industrial use.* BS1129, BSI, London, 1966.
12. BRITISH STANDARD CODE OF PRACTICE. *Metal scaffolding. Part 2: Suspended scaffolds.* CP97, BSI, London, 1970. New edition being drafted as BS5974.
13. NATIONAL ASSOCIATION OF SCAFFOLDING CONTRACTORS. *User's guide to temporary suspended access.* NASC, London, 1979.
14. PREFABRICATED ACCESS SCAFFOLD MANUFACTURERS ASSOCIATION. *Operators code of practice.* PASMA, 1980.